GRANDMA MADE YOU FAT
IT'S NOT YOUR FAULT

Thomas McGann

Agapi tis Glossas

ISBN: 978-0-9906241-6-5

LCCN: 2023917401

Published by Agapi tis Glossas Publishers, USA

938 Virginia St. #201, Dunedin, FL 34698

Cover Photo:

CC0 Public Domain - https://www.maxpixel.net/photo-4828422

For my mother

Irene Ann Gompers McGann

She taught me to shoot for the moon

To believe in myself and anything is possible

Indeed, it is

GRANDMA MADE YOU FAT - IT'S NOT YOUR FAULT

An Evolution Disputation

Evolution is a symphony of the most exquisite music heralding our most precious gift—life. It is operatic, soothing, and, yes, violent to the extreme. It has rhythm, harmony, melody, and even lyrics. Percussion maintains tempo; strings hum melodies; woodwinds breathe continuity; brass blows death in harmony with life.

Attend to the silences between the notes.

Composers note musical compositions in scales heard by their hearts, directed by their hands, and orchestrated by their minds. So do scientists.

One such scientist was Charles Darwin (1809-1882), naturalist, geologist, and evolutionary biologist (before that term ever existed). Darwin heard the symphony of evolution from afar and took a five year voyage around the world on the HMS Beagle to find its sources.

Upon return, it took him over twenty years to compose his work, and it might have taken even longer had he not received a paper from Alfred Russel Wallace outlining Wallace's theories on evolution which mirrored Darwin's own. They decided to present their works jointly, and so, the theory of the natural selection of species was born.

In 1859 Darwin published his book, *On the Origin of Species by Means of Natural Selection, or the Preservation of Favoured Races in the Struggle for Life*, to a combination of outrage and acclaim. His composition is beautiful and we still read its score, but, eventually new composers come along scribbling new orchestrations and the music changed, continuing to grow.

One oft-repeated chorus—random mutation—turns out to have been an all encompassing, but discordant, refrain. Scientists subsequent to Darwin wrapped his theory into an ever-evolving synthesis with random mutation its mainstay. Mutations, in the understanding of the times, meant permanent changes to DNA. This thesis assumed happenstance—of course—by chance.

However, random mutation turned out to be a just-so story, a fanciful narrative to explain unexplained evolutionary assumptions. Rudyard Kipling, the Nobel Prize winning author of such works as *The Jungle Book*, *Gunga Din*, and *Kim*, told his daughter Josephine bedtime stories that had to be worded "just-so," stories about how the leopard got its spots, about how the camel got his hump—*Just So Stories* with fanciful explanations.

One of Kipling's just-so stories tells how the elephant got his trunk, an elephant with a blackish, bulgy nose no longer than a boot, an elephant's child with 'satiable curiosity, who asked ever so many questions about everything he saw, or heard or felt, or smelt. He wanted to know what the crocodile ate for dinner, but everyone said "hush" in loud and dretful tones (for that is how they always talk) and they all spanked him for his 'satiable curiosity.

Finally, the Kolokolo bird told him, with a mournful cry, that he could find the crocodile down at the great, grey-green, greasy Limpopo River all set about with fever-trees, and, indeed, the 'satiable elephant-child did find a crocodile there and asked what the crocodile ate for dinner.

With a wink of an eye the crocodile asked the elephant-child to come closer to his musky, tusky mouth (for that's how he always talked). He caught the elephant-child by his little nose, which up to that very week, day, hour, minute had been no bigger than a boot. Both pulled, and pulled and pulled, and that little nose stretched and stretched and stretched till it became really, truly a trunk which finally popped out of the crocodile's mouth with a plop that could be heard up and down the Limpopo River.

And that, my friends, is how the elephant got his frisking and whisking trunk, according to Rudyard Kipling.

Just-so-evolutionary-biologists had to figure out how evolution created new species out of old ones. So they told a just-so story about the randomness of mutations, a story that seemed to fit though no one could prove it and few chose to attempt to disprove it (for that's how they always talked). As generations of scientists continued their experiments however, it soon became obvious that a more expansive explanation was necessary.

Even today many just-so-evolutionary-biologists continue to recite that same just-so-story in spite of recent papers that have determined that not all mutations are random. There is now proof that some mutations are directed, i.e. steered. Could, then, not *all* mutations be directed? The answer to this question has far-reaching consequences in many scientific—and other—disciplines.

Perhaps it may be time to listen to a new symphony, something radical perhaps—something like Stravinsky's controversial *Rite of Spring*, something in which the orchestra plays a new refrain—one where no mutations are random. Will the result be derisive laughter? Will there be riots? Will the thought police cancel crazed concepts? Let's take a listen.

THE DUTCH FAMINE BIRTH COHORT STUDY

It turns out that the blame falls on Grandma. Grandma made you fat. It's not your fault. That's right. Grandma is responsible for making you fat—as well as for causing your asthma, your diabetes, and your cardiovascular disease among other ailments. It doesn't matter whether your Grandma was the warm-huggy Aunt Bea type with an ample lap or the vile Cruella de Vil in her demonic Dalmatian wrap. It's all Grandmas' fault.

It turns out that what Grandma did during her lifetime, if dramatic enough, latched onto her genes and they got passed on to you. It may sound impossible, implausible, like passing the buck, but it's true. At least that's what scientists on the cutting edge of the revolutionary new field in biology called epigenetics claim.

And they can prove it.

Genetics is the study of genes, the study of the heredity of organisms. "Epi" is a Greek prefix meaning upon, over, on top of. So *epi*genetics is the study of what's happening upon, over, or on top of our genes. The term was coined by Dr. Conrad Waddington at the University of Edinburgh back in 1940. He used the word to explain the interaction between genes and the environment that he could not explain at that time using classical genetics.

The Dutch Famine Birth Cohort Study gave scientists their first inkling of Grandma's responsibility. During WWII, the western half of the Netherlands had been liberated by the Allies, but the other half remained in control of the Nazis. The Dutch government, in exile in London, ordered a national railroad strike to help disrupt the German occupation. In retaliation the Nazis ordered a food blockade in the entire occupied half of the country.

Prior to the blockade most Dutch were living on an adequate diet of roughly 2300-2500 calories per day. As food shortages began to take effect in October of 1944 the caloric intake dropped rapidly until May of 1945 when the Allies liberated the rest of the Netherlands. By then the daily ration was as low as 400-500 calories. There were an estimated 22,000 deaths by starvation and severe adverse health effects on untold others, not the least of whom were fetuses experiencing the famine in their mothers' wombs.[1]

Because of the specificity of the time period of the famine and the meticulous record keeping of the Dutch, scientists were able to study the effects of the famine on this particular population and on the children born immediately after its end. Studies began in the 1970s and continue into the new millennium.

During the famine Dutch women continued to get pregnant. Not surprisingly, the effects of the famine on the fetuses depended on the timing of their exposure to caloric deficiency. Women who had already gotten pregnant and then suffered through the famine when their fetuses were in the third trimester had babies who were more likely to be born small and with elevated levels of obesity, roughly double the levels of those born before or after the famine.[2]

On the other hand, babies who were conceived toward the end of the famine, those in their first trimester, who were starved at first and then well-fed afterwards, were more apt to be born at a normal weight, but they were also found to have higher incidences of obesity as well as a three-fold increase in cardiovascular disease later in life.[3]

Those babies, now adults, experienced famine related complications throughout their lives, problems such as higher levels of triglycerides, LDL cholesterol, and even schizophrenia. They died at a 10% higher rate than those born before or after the famine.[4]

And here's the kicker! When those children grew up and had children of their own, those offspring suffered consequences as well. The famine experienced by those children, now parents, caused changes that were passed down to their children *and* their children's children—transgenerational mutational changes. "[T]he increase in obesity, cardiovascular and metabolic disease previously described among the offspring of pregnant mothers exposed to famine during gestation is perpetuated in their grand-offspring."[5]

The celebrated Hollywood actress Audrey Hepburn was one of the victims of the Dutch famine. She and her family living in Arnhem, Netherlands, during WW II were reduced to making flour out of tulip bulbs to bake biscuits in order to survive and, as a result, Ms. Hepburn suffered severe health consequences for the rest of her life,[6] It is no surprise than that she dedicated her last years fighting famine as an ambassador for UNICEF.

Besides the Dutch exposure to the famine during WWII, a more recent study deals with the DNA signature left on the babies born after a 1992 Quebec ice storm.

Early in January 1992, a freak storm stalled over eastern Ontario, southwest Quebec Canada, and northern New York State resulting in heavy snow falls followed by a steep drop in temperature and freezing rain. Power was severely disrupted leaving more than 4 million people without electricity, most in the Canadian Maritime Provinces, some for as long as 45 days.

Following the storm, scientists initiated a study called Project Ice Storm examining the effects this storm had on pregnant women and their offspring. "Thirteen years later, the researchers found that DNA within the T cells [a type of immune system cell] of 36 children showed distinctive patterns in DNA methylation [the addition of a methyl group, CH_3]. The researchers concluded that maternal hardship is a predictor of the degree of methylation of DNA in the T cells. The 'epigenetic' signature plays a role in the way the genes express themselves."[7] These results were published in 2014, and the research continues on the children birthed following this storm.

OTHER STUDIES

Interest in epigenetics has intensified since the Dutch Cohort Project and Ice Storm studies and it is now a generally accepted discipline. Like a loose tooth to which the tongue constantly returns to wiggle and waggle, scientists revisit epigenetic transgenerational inheritance wiggling and waggling various life forms to see what they can learn.

- In plants – a "monster" epigenetic variant of the perennial toadflax [*Linaria vulgaris*] has been growing in the same region for over 200 years.[8]
- In bacteria – "[T]he French microbiologist Lisa Hirschbein and her colleagues found epigenetic inheritance in *Bacillus*...Daughter cells inherit and transmit this chromosome's inactive state for many generations."[9]
- In fruit flies – Researchers have found that by modifying certain proteins in *Drosophila* fruit flies they can alter the eye color of the flies without altering their DNA. They found that regulating gene position can lead to transgenerational inheritance of this trait.[10]
- In monarch butterflies – While studying how monarch butterflies (*Danaus plexippus*) came to eat toxic milkweed with impunity researchers at UC Berkeley studied the

effects of the toxin on the sodium/potassium pump cells used to balance the sodium/potassium gradient. This ability in insects resistant to this toxin required three coincidental identical mutations in a specific order. Armed with this knowledge the scientists concluded that "Because this trait has evolved so many times, we have also shown that this is not random."[11]

- In roundworms (nematodes) – Scientists introduced a gene into worms that made them glow in the dark under ultraviolet light. By varying the temperature in which the worms lived they could control the length of time during which the worms glowed and instill in them a memory of glowing brighter for longer periods. Furthermore, that memory was passed along for fourteen generations.[12]

- In addition, according to Columbia University's Oded Rechavi, evolution of acquired traits exists in at least one species of roundworms. "[R]oundworms that developed resistance to a virus were able to pass along the immunity to their progeny for many consecutive generations... [E]ven a hundred generations after the initial infection, the roundworms still carried the immunity acquired by their distant ancestor."[13]

- In mammals – male rats exposed to the fungicide vincloxolin in utero have reduced fertility as adults and their male offspring also have low fertility rates up to the fourth generation without further exposure to the fungicide. They also suffered diseases of the testis, prostate, kidney and immune systems. Other research projects involving agouti mice also demonstrated epigenetic inheritance and are among the best examples documented in mammals.[14]

- In humans – Barbara Wright and her colleagues at the University of Montana, Missoula successfully predicted directed mutations in three systems: *E. coli* (a gut bacterium) under phosphate starvation, oxidative stress and osmotic stress. They also found directed mutations in humans, in the tumor suppressor gene, and abnormally frequent mutations during the immune response to foreign antigens (substances alien to the body that evoke immune responses).[15]

So, what's happening?

Forces in their environments stress organisms. If the stress is severe enough over a long enough time period the organism reacts by attaching methyl groups near, upon, on top of its genes, i.e. some DNA is methylated. This means that a small methyl group (three hydrogen atoms attached to a carbon atom, CH_3) is attached to some of our genes. These are epigenetic changes—modifications to genes that change the way genes are expressed but which did not alter the genes themselves. Think of it as a kind of volume control that turns up the loudness of some genes while suppressing it in others.

JEAN-BAPTISTE LAMARCK

Jean-Baptiste Lamarck by Charles Thevenin - Wikipedia

Much of the skepticism voiced in opposition to epigenetic transgenerational evolution is based on the long-ago widely discredited theory of the inheritance of acquired characteristics (also known as soft inheritance) that Jean-Baptiste Lamarck espoused early in the nineteenth century, a theory which many scientists have no desire to resurrect or even discuss resurrecting. It is a battle won long ago—or so they argue

Jean-Baptiste Lamarck (1744 –1829) was a French naturalist/biologist and author of the 1809 book *Philosophie Zoologique ("Zoological Philosophy, or Exposition with Regard to the Natural History of Animals")* which included the theory of the inheritance of acquired characteristics—the idea that living creatures pass on to their offspring characteristics that they have acquired during their lifetime through either use or disuse. For example, giraffes have long necks because they stretched their necks to feed on the tender shoots on tree tops. Their offspring would inherit long necks, feed on tree-top shoots and further elongate the neck. Voila—today's giraffes!

Giraffe on the African savannah

With the advent of Charles Darwin, Lamarck's theory was dumped into the dust bin of history. Lamarckism is so discredited today that its mere presence in scientific literature is usually mocked. However, new studies into epigenetics are giving even those most opposed to Lamarck's idea, pause to reconsider.

CHARLES DARWIN

Charles Darwin by Julia Margaret Cameron - Wikipedia

Fifty years later, along came Charles Darwin author of *On the Origin of Species.* In it Darwin proposed the theory of survival of the fittest by natural selection. All organisms are descended from a single common ancestor, LUCA (Last Universal Common Ancestor). Each species has developed its own variations from the others enabling that species to be better—or worse—at surviving and reproducing. Beneficial mutations foster the survival of species while deleterious mutations eventually end in extinction.

Here is a brief summary of Darwin's Theory of Evolution:

- A species is a population of organisms that interbreeds and has fertile offspring.
- Living organisms have descended with modifications from species that existed before them.

- Natural selection explains how this evolution has happened.
    - More organisms are produced than can survive because of limited resources.
    - Organisms struggle for the necessities of life, i.e. there is competition for resources.
    - Individuals within a population vary in their traits. Some of these traits are heritable and are passed on to offspring.
    - Some variant traits are better adapted to survive and reproduce in a given environment than others.
    - Better-adapted individuals (the "fit enough") are more likely to survive and reproduce, thereby passing on copies of their genes to the next generation.
    - Species whose individuals are best adapted survive—those who are not become extinct.[16]

Darwin used the term "natural selection" as opposed to "artificial selection" to specify the exact mechanism driving survival. Artificial selection (also called selective breeding) is the calculated choice of desirable traits by human intervention—something that we have been doing for centuries. This was a major source in furthering Darwin's understanding of speciation, i.e. how one species develops into another.

Artificial selection amply demonstrates the effects of the environment on changing the phenotype (the observable characteristics of individuals) of domesticated plants and animals. ***We*** become the environment. We have cultivated broccoli, cauliflower, kale etc. from the wild mustard plant. We have bred dogs from wolves into breeds as varied in size as Great Danes to Chihuahuas, and as varied in coats as red Irish Setters to mop-coated Komondors.

Natural selection, on the other hand, occurs in the wild—in nature. Both methods select for desirable traits, one by man the other by the environment. Natural selection determines which changes to individual phenotypes are inherited through time, which small variations increase the individual's chances to survive and reproduce. It is also called "descent with modification" or "survival of the fittest," where some traits are favored by the environment while others are not. Natural Selection changes species in two ways, microevolution which changes color or size (for example), and macroevolution which changes one species into another (speciation). Natural selection is the mechanism driving evolution. Darwin himself put it this way, "Natural Selection...is a power incessantly ready for action."

In his *The Origin of Species* Darwin wrote: "I have hitherto sometimes spoken as if the variations … were due to chance. This, of course, is a wholly incorrect expression, but it serves to acknowledge plainly our ignorance of the cause of each particular variation. [The facts] lead to the conclusion that variability is generally related to the conditions of life to which each species has been exposed during several successive generations."

Things that make you go hmmm...

BIOLOGY OF EPIGENETICS

Do not panic. For those of you, whose last exposure to biology was your sophomore year of high school, stick with me here. It's not complicated.

DNA is a molecule shaped like a free-form spiral staircase (or twisted ladder), a double helix that coils through space without need for a central post. That DNA contains the genetic instructions for all living organisms. The stringers for our spiral staircase, those supports on both sides of the stairway, are its sugar-phosphate backbones and are joined together by stair treads (or ladder rungs) called base-pairs.

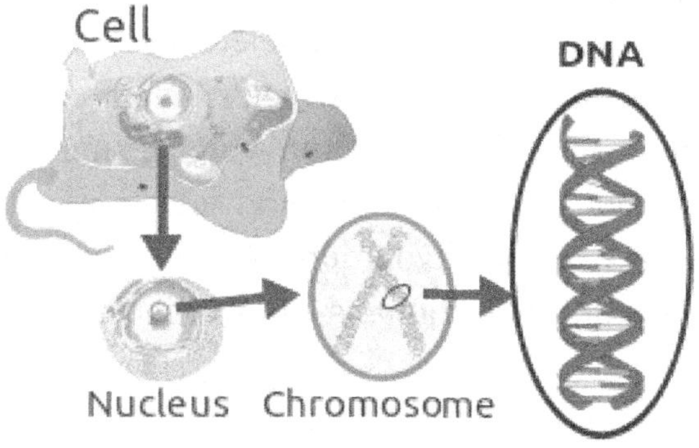

Organization of DNA in a eukaryotic cell
Creativecommons.org/licenses/by-sa/3.0/deed.en

Some of our DNA can be methylated. This means that a small methyl group (CH_3 - one carbon atom bonded to three hydrogen atoms) is attached to parts of our genes. Methylation determines whether these parts, called alleles, are turned on or off—how they are expressed. Varying DNA methylation patterns affect the how, when and where of gene activation.

Genes are sections of the DNA spiral like a section of the staircase between the floors of a building. Alleles are alternative variations of genes; different design features that code for different physical traits such as eye color, height, etc. It is the allele that becomes methylated. In addition to alleles new advances in cell biology have added additional inheritance mechanisms: histone modifications, chromatin structure, and a select group of non-coding RNA.

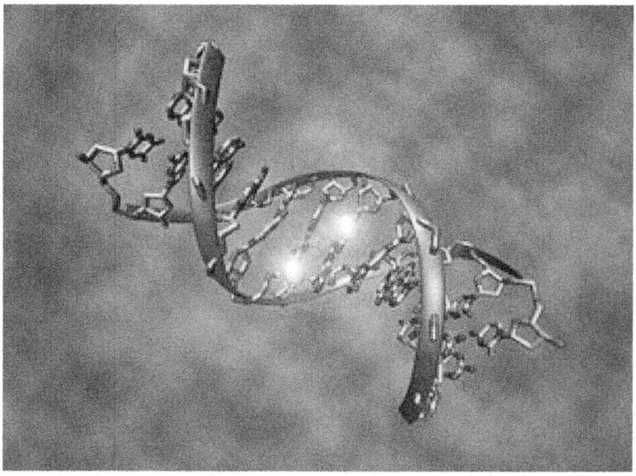

Representation of a DNA molecule that is methylated. The two white spheres represent methyl groups. They are bound to two cytosine nucleotide molecules that make up the DNA sequence.
Creative Commons – Christopher Bock, Max Planck institute.

These mechanisms also transmit this epigenetic information from one generation to the next. "The effect of methylation is to inhibit the expression of the gene to which it is attached…methylation is not transient; the methyl group tends to stay attached to the DNA, even after it replicates during cell division. The methylated DNA not only persists throughout the life of the cell but is transmitted to all of the cells that descended from the cell in which the original epigenetic change occurred. As such, those genes that are epigenetically turned off as a result of methylation tend to stay turned off in that cell lineage.

"But at certain critical periods in early development, things are in flux, methylation-wise. Some biochemical pathways promote methylation, and other biochemical pathways prevent methylation or even cause demethylation."[17] [Removal of methyl groups.]

Think of methylation as adding chocolate to milk. Chocolate changes the flavor as well as the color of the milk without changing the milk itself. Nestlé's Quik adds a different flavor than U-Bet or Hersey's syrups, each without changing the structure of the milk.

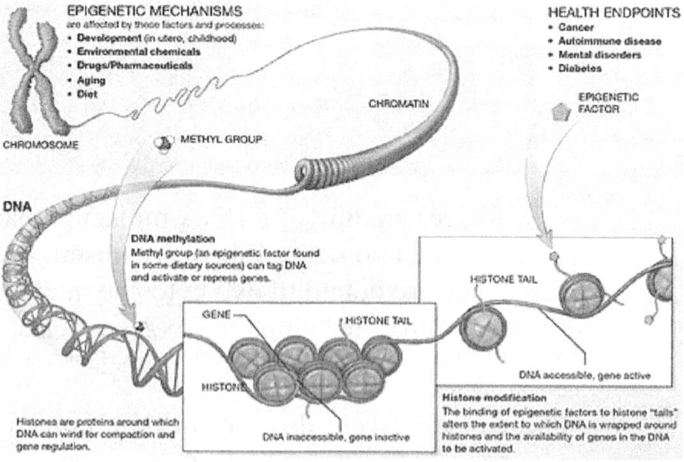

By Christoph Bock, Max Planck Institute for Informatics - Own work, CC BY-SA 3.0,

The most numerous documented examples of epigenetic inheritance occur in plants and fungi. One supposition is that plants and fungi are more adaptive to epigenetic inheritance than animals because plant/fungi environments are more constant. Any environmental tags to which they are exposed would be repetitive year after year and would be more likely to stick to the genes. Animals, being more mobile, are exposed to more variable environments with the likelihood of less of a continued exposure to the same tag frequency. These inheritances involve speciation, the formation of new species. Darwin emphasized that "species is a fundamental category of biological organization."

SPECIATION

In the mid-eighteenth century a Swedish botanist, zoologist and physician by the name of Carl Linnaeus (1707-1778) published a number of books that began classifying plants and animals in Latin with a binary nomenclature, e.g. *Homo sapiens* (that's us). Other examples are: *Ursus horrobilis* (grizzly bear), *Bison bison bison* (American plains bison). Some are named after famous people. For example, Keith Richards (*Perirehaedulus richardsi) and* Mick Jagger (*Aegrotocatellus jaggeri)* have (aptly) fossil trilobites named after them. Trilobites are now extinct arthropods (spiders, and crabs.)

Linnaeus' work has become the basis for the binomial system of naming organisms that we still use today. We have learned much by our recent developments into DNA sequencing that has helped establish the evolutionary relationships between the various life forms but Linnaeus' system remains the standard.

Today we recognize seven tiers of biological classification we call taxa (singular is taxon). They are, in order from the most general to the most specific: kingdom, phylum, class, order, family, genus and species. The second taxon is divided into two parts, one called *phylum* for animals, and the other called *division* for plants. There are numerous terms for naming genetic variations below the species level, but that naming gets murky as different groups use differing words. There are subspecies, breeds, varieties, hybrids, strains, cultivars, and landraces.

Confusing? You bet. Take the common tomato for example. There are over 10,000 tomato cultivars, (cultivated varieties), the work of human hands. But is the tomato a vegetable or a fruit? Genetically, the tomato is a fruit, a berry of the plant *Solanum lycopersicum*, a ripened flower ovary with seeds, but to nutritionists the tomato is a vegetable. There was even a court case on this very matter that went all the way up to the U.S. Supreme Court, *Nix vs. Hedden*, in which the tomato was deemed—wait for it—a *vegetable* by a *unanimous* vote of the court. There is an old adage that goes: "Knowledge is knowing that a tomato is a fruit; wisdom is not putting it in a fruit salad."

All this attention is due for the need to protect those in the business of cultivating new varieties of plants and animals, but in discussions of evolutionary biology we are interested in where on the tree of life species and their subunits—whatever they are called—fall.

Consider the myriad forms of life here on Earth: the paramecium and the tardigrade; the Venus flytrap and the white dove orchid; the hissing cockroach and the glass-wing butterfly; the duck-billed platypus and the bottle-nose dolphin; etc.

Why so many and why so varied?

Speciation is the process by which one population of organisms (a species) evolves into a new and distinctly different species. But, why species? Why not just one life form? If life needs to live why not foster a single *super* species? Perhaps, the reason for the existence of so many varied species could be a need to expand the possibilities for life to exist in whatever environment it finds itself.

Whatever the reason, we find ourselves thronged with species. These initial self-replicating organisms exposed to their varied environments drifted apart genetically, over eons, creating this vast variety of life forms both extinct and extant.

Currently, some scientists object to the use of the word species since it cannot be explicitly defined, because there are exceptions to each definition—in particular, asexuality (no fusion of sperm and egg), and hybridization (crossbreeding between subspecies).

Of the numerous definitions of species available, in this discussion the one we will be using is that of Ernst Mayr (1904-2005), one of the 20th century's leading evolutionary biologists. He defines species as "groups of actually or potentially interbreeding natural populations, which are reproductively isolated from other such groups." Reproductive isolation is the key concept here, meaning that one species is unable, for whatever reason, to produce fertile offspring after mating with a member of a different species.

Examples of two species that originated from a common ancestor but cannot produce fertile progeny are the horse and the donkey. The horse *(Equus ferus)* and the wild African ass (*Equus africanus*) both evolved from a species called *Dinohippus* some 6 million years ago (mya), with the separation of the two species occurring approximately 4 mya. The donkey/ass (*Equus africanus asinus*) descended from the wild African ass and was domesticated about 3000 BCE. Today, when the ass and the horse breed the offspring is almost always sterile. The results are mules and hinnies. The reason horse/donkey hybrids are sterile is because horses have 64 chromosomes while donkeys have only 62. Perhaps, given enough evolutionary time, they will no longer be able to birth any offspring, sterile or not.

When populations of species are geographically isolated from one another, separated into two or more groups subject to different environmental pressures, they are imprinted by the attachment of epigenetics markers to alleles of their DNA. Natural selection then winnows these groups on the basis of their ability to survive and reproduce. Eventually this process mutates the original species into an entirely new species, two that are no longer able to mate and produce fertile offspring. Note that until this stage is reached the two groups can still mate and mix genes, piggybacking epigenetic markers. This is called hybridization.

There are four types of speciation: allopatric, peripatric, parapatric, and sympatric. Basically, all four are differing descriptions of similar events—populations of species getting separated from one another for varying reasons. Because of their differing environments they begin to drift apart genetically (genetic drift) as their epimutations begin to express and/or suppress their alleles. Given enough evolutionary time new species may be created. Until then they are subspecies, varieties, cultivars etc.

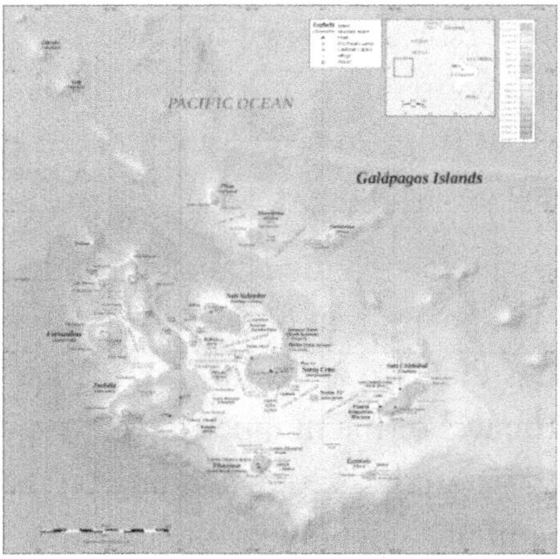

For examples of speciation let's return to Charles Darwin's visit to the Galapagos Islands on the HMS Beagle. The Galapagos Islands consist of some 19 islands, 13 large and 6 small, covering an area of over 17,000 square miles. Ecuador, South America, is their closest landmass, some 600 miles distant. The islands themselves extend about 160 miles east to west and 150 miles north to south. Some are close to one another while others are more distant. It was here that Darwin began his studies of finches.

The commonly accepted theory is that the first finches—actually some sort of blackbird or mockingbird—arrived on the Galapagos Islands from South America a few million years ago, established themselves, and started breeding. Because of the distance from South America, and the distances between the various islands, new species were generated over time by speciation, i.e. isolated groups on different islands bred within the confines of their new environments creating new species of finches. The process by which species spread to empty niches is called adaptive radiation (radiating out from an origin point and then adapting to their new environment).

Though Darwin was on the islands for only five weeks, among the 14 species that he found, he discovered 12 brand new species that had evolved from the ancient ancestral birds [estimates of the number of finch species on the Galapagos vary to a high of 20 depending on the scientist].Though similar to mainland finches, they varied in body size, beak shape, song, and feeding patterns. Of the species Darwin found, three were ground-dwelling seed eaters, three lived on cactus and survived on its nectar, one lived in trees (also eating seeds) and seven were tree-dwelling surviving on insects. Each species evolved beak types and sizes best suited for its most plentiful food source.

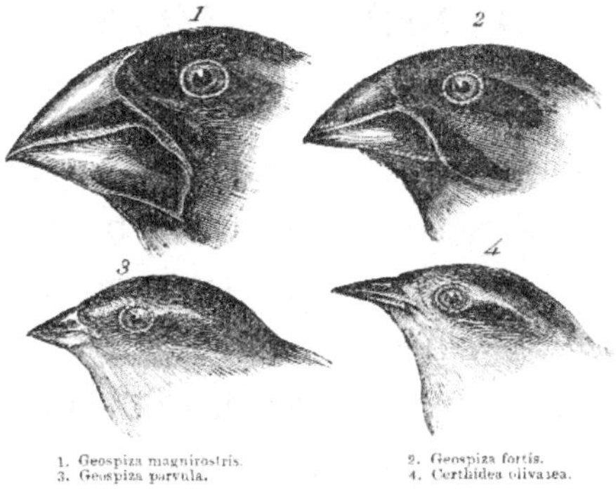

Darwin's Finches - From South America to the Galapagos Islands
Public Domain

Studies into speciation continue. Recent work (3/18/2020) by Laura van Holstein at St. John's College, University of Cambridge published in the Proceedings of the Royal Society, Biological Sciences, has confirmed the Darwinian hypothesis that in species rich in genera (biological rank as previously mentioned) there should also be subspecies richness. To quote Darwin, "Where many large trees grow, we expect to find saplings." Van Hosstein's research verified Darwin's expectation. She also considered whether the evolution of subspecies could be a step toward the creation of new species and found that they are, indeed, an early stage of speciation.[18]

There is hope for establishing clarification for the meaning of species. Given the ongoing advances in genome sequencing it may be possible in the near future to delineate species entirely by their chromosomal makeup and we will finally get a decisive definition of "species," one in accord with Mayr's theory of reproductive isolation.

HYBRIDIZATION

Hybridization occurs when two "species" interbreed creating a hybrid "subspecies". These words are placed in quotation marks because the distinction between the two is still in debate. In the case of the Galapagos Island finches, there are "species" inhabiting various islands that interbreed producing hybrids. Some of the times these hybrids who interbreed continue to be fertile.

The reproductive isolation here is more than just geographical and biological however. It is cultural and/or morphological as well (e.g., the members of one hybrid group do not find the song or the beak size of competing suitors attractive enough to breed with them, and, hence, are reproductively isolated.) Should environmental conditions change sufficiently, should there be a dearth of available males or females with the correct song and/or a big enough beak, this cultural/morphological reproductive isolation may cease to exist and they may become less choosey in their selection of mates.

In 2018 the team of Rosemary and Peter Grant, married naturalists who have been studying Darwin's finches on the islands for more than forty years, determined that rapid speciation can occur in as little as three generations. They followed a rare colonization event when a "big" male migrant finch, *Geospuza conironstris,* arrived on the island of Daphne from its usual habitat on Espanola, some 100 km distant. He mated with a female *G. fortis* and by 2012 there were 36 individuals and 8 breeding pairs of hybrids. The Grants called this their "Big Bird" lineage because of the size of the original male.

The Grants do not specify a definition of species in the literature I have found. They base their classification of the "Big Bird" lineage on the reproductive isolation of the family, not on the biological definition of species as defined by Mayr. Given that many of the populations of hybrid finches on the islands continue to mate and are able to produce fertile offspring, a logical inference would be that the island's finches consist of a continuum of varieties (subspecies) of hybridizing finches. Given another million years (or so) perhaps new species will evolve that will then be unable to mate and reproduce. What is occurring on the Galapagos Island among its finches is hybridization on its long journey toward the formations of new species.

RARE HUMAN EPIMUTATIONS

Just like the tomato, *Homo sapiens* also have epigenetic varieties. Examples of some rare human variations are:

- One such example is the high-altitude populations of humans who live in the Tibetan (Asia)/Andes (South America)/Ethiopian highlands (/Africa). Given the low levels of oxygen at the high altitudes in which these populations live, the stresses on human physiology are extreme. At an altitude of 9000 ft. the oxygen level is approximately 30% less than at sea level. Consequently, these populations have developed an ability to deliver small amounts of oxygen to their lungs enabling their bodies to operate efficiently at high altitudes. This ability is the result of an epimutation of a gene called *EPAS1* which regulates the production of hemoglobin.[19] Not surprisingly, women with this epimutation have a higher success rate of children who survive than do those without this adaptation.[20]

The Ethiopian highland town of Bekoji is an example of the result of this mutation. The town, home to only 16,000 individuals, is 2810 meters (just over 9,000 feet) above sea level. Yet this small town has produced an incredible eight Olympic gold medals, 10 world records and 32 world championships in long distance track and field competitions.[21]

- Another example is the Banjau of the South Seas who live entirely at sea and have an increased ability to remain underwater for lengthy periods of time due to the increase in the size of their spleens, (genetic variant to the *PDE10A* gene).[22] One report claims a Banjau dove to a depth of 230 feet and could remain under water for up to 13 minutes.[23]
- Other rare human epimutations include resistance to extreme cold in the Inuits, indigenous Siberians, and aboriginal Australian populations, (mutation of gene *CPT1A*).[24]
- In another, the need for less sleep (mutation of the gene *DEC2*).[25]
- Also, on the continent of Africa where malaria is prevalent some Africans have developed a resistance to malaria because of a

sickle-cell mutation (a mutation in the hemoglobin-Beta gene found on chromosome eleven).[26] Between 10-40 percent of people living in these areas carry this variation helping them survive malaria. Unfortunately, if this mutation is inherited from both parents their children become victims of sickle-cell anemia which can be fatal.

These adaptations indicate that human culture and biology have been co-evolving for millennia.

ASEXUALITY

Why sex?

Cloning (asexuality) would seem to be the easiest method of reproduction. It requires the least output of energy, whereas with meioses (sex between females and males) it takes two to tango. Sex requires the production of sperm and egg, finding a partner, and the risk of mating that could result in physical harm or infections. Just as light passing through changing mediums "selects" the "laziest" route, the one requiring the least time [more on this later], so too does asexuality "choose" the "laziest" route for procreation. Why bother with sex?

Meiosis as we humans know it, non-clonal sex, is nearly universal and it differs depending on the organisms involved—a food-court of tastes if you will—from birds and bees, to seahorses (where the males get pregnant) and scorpion flies where the female mates with more than one male and the female then decides who the father will be depending on the size of the nuptial gift.

Sex must be of paramount importance among all species otherwise they would all go extinct. The major benefit of meiosis is the coming together, the recombination, of two different chromosomes which generates genetic diversity.

Additionally, there is the benefit of the ability to repair DNA. If DNA is badly damaged, a diploid organism (two sets of chromosomes) has a second copy, a template, from which it can copy itself and make accurate repairs. This ability is reduced in haploid cells because only one chromosome is present.

Another possible answer as to why meiosis sex evolved is stress. In studies conducted by Aurora Nedelcu of the University of New Brunswick on algae she found that the algae live an asexual life until stressed (too hot, lack of nitrogen, etc.) at which time they resort to meiosis sex. The algae form diploid cells to enable them to survive hostile environments. When conditions improve they return to their haploid, asexual, state.[27]

"Theoretical studies of the evolutionary consequences of losing sex (aka the evolutionary advantage of sex) have shown that asexual organisms should accumulate more detrimental mutations than do sexual species and in consequence should have a higher rate of extinction. Also asexual organisms should be less able to fix advantageous mutations, which can also contribute to extinction in a changing environment, as well as making it more difficult to speciate by adapting to different niches.

"...In most asexual organisms, including land plants, there appears to have been a single ancestor from which they originated and their diversification appears to be comparatively recent leaving insufficient time to diverge into distinct species. In fact, the very existence of speciation in asexual organisms remains in question (Maynard Smith and Szathmary), but a recent study may prove otherwise as it has shown that asexual organisms can form species as defined by a number of shared species concepts."[28]

In an experiment recently published in *Cell Reports* (10/2020), researchers manipulated individual cells of baker's/brewer's yeast, *Sacharomyces cerevisiae*, which reproduce asexually about every 100 minutes. They found that the reduced gene expressions could not have resulted from gene expression alone. Epigenetic causes must have been involved.[29]

Lamarckian evolution raises its unheralded head once again.

[For an in depth examination of speciation in asexual organisms see *Lost Sex The Evolutionary Biology of Parthenogenesis* Edited by Peter Van Dijk, Koen Martens, Isa Schön.]

GENETIC DRIFT

Genetic drift happens when a segment of a species becomes isolated from its mother group and is exposed to a different environment. This change creates a change in allele frequencies in populations from generation to generation such as with the Galapagos finches.

Genetic drift can occur because of chance events such as earthquakes, typhoons, volcanic explosions, and the like, which wipe out substantial segments of populations of organisms. These events are called population bottlenecks and the results can be dire. Although these cataclysmic accidents happen in populations of all sizes, their effects are exaggerated in smaller populations. If a volcanic explosion wiped out a large segment of a species, the odds of survival of the remaining members would be severely reduced. If it occurs in a small population the results would be even more greatly enhance. An even smaller population would survive creating an even smaller gene pool from to which to draw. This kind of genetic drift can cause genes to disappear completely thereby reducing genetic variation, but it can also cause rare alleles to become more frequent, even fixed.[29]

PLASTICITY

There are segments of DNA that are called "hotspots," methylated CpG islands. They are called "hotspots" because mutations occur at these islands in greater frequency than other spots on the DNA.[30] This is a nonrandom pattern, occurring in hotspot clusters, caused by predation, competition, and/or other stresses.

Plasticity means changes in the phenotype that are reversible. As the environment changes the organism may change with it. Plasticity is adaptation by the alleles that enhances survival and reproductive success of the phenotype, but which is readily able to revert to its original form. If and when an environment changes any phenotypic changes may be removed and replaced by others. The very term "plasticity" reflects this transient nature—here today, gone tomorrow—*maybe.* "[N]atural selection is not always going for the optimal solution because the optimal solution is changing all the time. What is best for natural selection in the short run may be the wrong selection tomorrow."[31]

In a paper dated 22 May 2020, scientists studied the plasticity of Tibetan chickens to adapt to their ancestral environments and then back again. Their experiments also included guppies and the rod-shaped bacterium *Escherichia coli.* The results demonstrated that organisms remember their ancestral environments and adapt accordingly. They undergo changes to their epimutations without permanent mutational changes to their DNA.[32]

Genetic drift and plasticity are terms describing similar processes. One drifts and is established; the other drifts and comes back.

This is analogous to neuroplasticity which is the ability of neural networks in the brain to respond to environmental pressures by rewiring themselves, establishing new synaptic connections to nerve cells other than those to which they are usually connected. Some neural networks appear to be hardwired while others are able to deviate from their normal functions to perform new objectives while also retaining the ability to revert to their original functions in a constant wiring/rewiring scenario during which the brain continues learning. Research has determined that the brain continues to change its architecture throughout our lives, reorganizing connections, enabling us to respond, to exhibit behavioral flexibility.

MUTATION/EPIMUTATION

According to Darwinian evolutionary theory, all species evolved from a single organism. Following the inception of that organism, new species emerged by incremental changes to DNA over evolutionary time, millions and millions of generations. But how did these changes occur? What mechanism drove them?

Darwin proposed that this was due to the struggle for life, the drive to survive, with the winners determined by natural selection. He had no concept of genetics which entered scientific literature with the rediscovery of the work of the Augustinian monk Gregor Mendel in 1900.

According to the thinking during those times, Darwin, and subsequent scientists, assumed that the mechanism used by evolution was chance, random mutation. Without mutation by chance there would have had to have been a prime mover, some primary cause, some kind of cosmic purpose, an omega point, a "god".

The assumption of randomness was in direct refutation to the prevailing idea of the time, creationism—that is that all life was a result of the handiwork of some godhead. However, the "god" idea was not tenable. Evolution is science not theology.

Unable to explain the origin of mutations, scientists made the assumption that mutational changes occurred by chance. They fell back on a "no argument" argument, the "just so" story because it made sense, of sorts, even though it could be neither proven nor disproven. It could not be falsified either.

The assumption that mutations occur randomly is just that, an assumption without any scientific basis. The theory of parsimony/Occam's Razor, that the simplest of explanations is most likely correct, seemed to be the earliest, the easiest, the most natural assumption, the first grasp at the handle of understanding of how natural selection works. However recent studies have shown us that not all mutations are random, that some, at least, are directed by the environment.

Karl Popper (1902-1994), one of the most renowned philosophers of science, is famous for the doctrine of falsifiability. A hypothesis is falsifiable, he argued, if it can be proven to be false by observation. The assertion that all mutation is random is unfalsifiable on its face because proof could only be obtained by observing every mutation that ever existed and comparing them to previous mutations looking for connections of descendency. If, however, we turn this thinking on its head, and prove that one mutation is not random, that it is directed, we must conclude, at the very least, that not *all* mutations are random.

The claim "all swans are white," once accepted as fact, was falsified when, in 1697, the Dutch explorer Willem de Vlamingh sighted black swans on a river near Perth, Australia. The moment Willem first saw black swans was so amazing that it has been immortalized with a statue erected near that site of a wide-eyed Willem spotting a nearby black swan, wings extended, ready to take flight.

<img src="images/vlamingh-swan-river.jpg" alt="" />

All swans are white except those that are not. All mutations are random except those that are not.

Given the successful history of an organism's development from its first life form to its present embodiment and all the steps in between that have already been programmed into the organism, could the mutations be "best guesses" as to what will work as opposed to those that will not? If these mutations result from eons of classical evolution that has led to the organism's successful survival to date, are they then not more likely to produce more successful mutations than mere random, mere chance events? Is not the direction of success already preprogrammed?

Some of these inbred mutations have been around for so long that we have named them instincts—behavior driven, seemingly without cause. Instincts, the oldest memories buried in the early vestiges of a species—how species came to be themselves—are carried forward through time.

"The longer the existence of an instinct, the stronger is its hold on the evolution of that species and the more resistant it is to change. The DNA replication fidelity is so great that relatives of many human genes can be identified in yeast, showing that some DNA sequences from the single-celled common ancestors of humans and yeast have been transmitted across generations with little change for well over a billion years."[33]

A current, well-accepted theory is that most mutations are deleterious. But are they? Perhaps most mutations are merely imperfect in their ability to perfect the necessary answers to the problems posed by the environment and, as so, are classified as deleterious when, in reality, they are nothing more than imperfect answers to the problems posed.

Some DNA sequences may be incorrectly copied and a permanent change in the code of DNA, a mutation, may occur. These mutations are merely mistakes, however inaccurate, in the copying or repair processes in which they were involved. As such they are not random. They are mistakes, perhaps even reasonable facsimiles, but not random. Attempting to repair a screen door that fails to latch properly, a new latch is fashioned similar to the original but different and it may work just fine. If the repair does not work, if the latch fails to catch, the screen door falls short of its purpose. Those mutations die.

We have learned that directed mutations are caused by the methylation of alleles, histone modifications, chromatin structure and a select group of noncoding RNA driven by forces of the environment. As such they are called *epi*mutations, and these epimutations can become imprinted and inherited across generations. But, it is argued, they are not "real" mutations because there is no change to the organisms DNA. They can arise quickly and just as quickly "sunset" giving organisms the ability to weather temporary stresses of their environments (plasticity). Given forces of sufficient magnitude over evolutionary time, these epimutations could eventually facilitate fixation, could become DNA permanent mutations as opposed to epimutations.

In 2015, in an article entitled "Environmentally induced epigenetic transgenerational inheritance of sperm epimutations promote genetic mutations,"[34] the authors investigated germline (egg and sperm) transmission of epigenetic information between generations. Their study "was designed to investigate the influence of the epigenetic transgenerational inheritance of germline epimutations on the occurrence of genetic mutations using CNV."[CNV – Copy Number Variations].[35] "The hypothesis tested is that developmental exposure to environmental factors (e.g. the toxicant vinclozolin) can promote the epigenetic transgenerational inheritance of germline epimutations that influences genome instability and genetic mutations (e.g., CNV). Therefore, the transgenerational phenomenon is initially induced through epigenetic inheritance but, in later generations, the transgenerational phenotype may involve a combination of the effects of epimutations and derived genetic mutations."[36]

The results of this study found that epimutations lead to permanent mutations. Evidence of effects of the environment on epigenetic transgenerational inheritance appears to be an example of epigenetics driving genetics. Additional research has shown that copy number variations are significantly more common in genes that encode for proteins that directly interact with the environment than proteins that are involved in basic cellular activities.[37]

So we've come full circle. The environment affects the epigenetics of the DNA which, in turn, mutates the DNA given sufficient force and time.

xxxSo, are no mutations are random? Are all driven by the environment?

RANDOMNESS

Randomness is hypothetical – a just-so-story.

Something is random by virtue of its deviation from an anticipated pattern, the deviation from the pattern that determines what is random and what is not. Consider two concepts of randomness:

- Concept 1 – An event is assumed random due to our ignorance of its causes.
- Concept 2 – An event is said to be random when it is the result of independent causal chains—consider traffic accidents.

But, perhaps both concepts, 1 and 2, are the same—the only difference being our lack of knowledge of all the factors involved in the causal chain of events, our ignorance of what actually caused a particular traffic accident? If a car involved in an accident left one second earlier or later would the accident never happen? How many factors do we need to know to explain why the accident was not random? There are the obvious ones such as route chosen, speed, amount of traffic, weather, etc, but how about the less obvious such as tire pressure, texture/pitch of the roadway, alertness of the driver?

There is another possible definition of randomness. We could define a random/spontaneous event as one that occurs so infrequently as to be called random, but that just confuses both the logic and definition of the word, confusing matters even more. Infrequency separates the causes of an "accident" over long periods of time making determination of its causes even more difficult to determine.

Let's take a philosophical run at randomness in evolution. How could we go about proving that mutations are random? Evolution has been at work for thousands of millions of years, efficiency being an apparent, but an, as yet, undeciphered goal with life as the prize.

When running a computer program using numbers that are rounded off, differing results will be obtained depending on where and how the rounding off takes place. The common method of rounding off is to "round up" if the decimal is above 5 and to "round down" if below 5. Errors will occur because of the rounding. For example, if an apple costs $2.49 we round that off to $2. If the same apple costs $2.51 we round that off to $3. So does a dozen apples costs $24 or $36?

Take pi, **π**, the ratio of the circumference of a circle to its diameter. Where do we round pi off if, for example, using it in calculating an algorithm? Pi is an infinitely long number of digits. Do we round it off after 3.14159 or after 3.14159265358979323...? With evolution at work on cell division and replication on an evolutionary time scale, how exact would cell replication have to be to create exact copies each time and not be random or spontaneous?

An explanation of the "how" of random mutation is lacking in scientific literature. Does a mutation occur all at once in numerous individuals or does it occur only in a select few here and there? It could occur all at once in many if the many were all exposed to the same stresses. If it appears only in a few the spread of that mutation would be slow.

Scientists insist that the frequency of spontaneous, i.e. random, mutations is rare. The probability of the occurrence of a spontaneous mutation is less than one-one-thousandth of that of an epigenetic change.[38] This creates a problem for the speed of evolution. How could all these evolutionary changes occur at so slow a pace?

Each iteration of cell division is an attempt by a life force to adjust to the ever changing environmental pressures in order to survive and procreate. Our inability to see all causes of mutations does not mean that the mutations are random.

ON SPONTANEITY

Recently some scientists have taken to calling random mutations *spontaneous* mutations. Spontaneous mutation, spontaneous combustion, spontaneous crystallization are processes that occur without the addition of a catalyst. A pile of oil soaked rags sitting in a hot garage, for example, ignites on its own—spontaneous combustion. If we remove any one of the three necessities for this to occur, heat, oxygen or fuel, the event cannot happen.

So, is any event ever really spontaneous? When Krakatoa erupted in 1883, or more recently Mount St. Helens in 1980, was either event spontaneous? Were either the Alaskan earthquake of 1964 or the Fukiashima quake of 2011 spontaneous, or were they were all the results of physical forces at work over time that finally expressed themselves in such cataclysmic fashions? If traced to their ultimate causes these events and others like them that we chalk up to spontaneity, to chance, are merely physics at work. If there is a reason behind an event, it is not truly spontaneous, not random. The forces within the volcanoes built up to such an extent that the explosions were inevitable. Same for the earthquakes. The tectonic plates needed to move and they did.

LIFE FORCE/ÉLAN VITAL/WILL TO LIVE/VITALISM

The spark of life is real!

In humans, photographs taken the instant sperm penetrates the egg, display a visible spark. Life begins with a flash of lightning![39] Google the footnote and watch it for yourself. It's amazing.

Although scientists squabble about its details, life is generally defined as incorporating the following characteristics:

- All life forms are made up of cells
- Maintenance of homeostasis, i.e. maintaining stability in its internal environment
- Metabolization, i.e. convert food to energy
- Grow and develop
- Reproduce and pass on genes
- Respond to the environment
- Death

Jacques Monod (1910-1976) the Nobel Prize winning biochemist in his book *Chance and Necessity* explains what is meant by life.

> "[F]rom the bacterium to man the chemical machinery [of life] is essentially the same in both its structure and its functioning...
>
> "In its structure: all living beings, without exception, are made up of the same two principal classes of macromolecular components: proteins and nucleic acids...and are constituted by the assembling of the same residues, finite in number: twenty amino acids for the proteins and four kinds of nucleotides for the nucleic acids...

> "In its functioning: the same reactions, or rather sequences of reactions, are used in all organisms for the essential chemical operations: the mobilization and storing of chemical potential, the biosynthesis of cellular components...
>
> "[Furthermore] with each succeeding cellular generation it is the *ne varietur* [it must not be changed] reproduction of the text, written in the form of DNA nucleotide sequences, that guarantees the invariance of species. The fundamental biological invariant is DNA."[40]

Life is *conditioned* to survive. It fights to survive. This fight goes on all around us, from the crashing heads of horny male big horn mountain sheep that can be heard from miles away, to the poisoned-tipped sword fights between bacteria.[41] The struggle to survive, to evolve in order to survive and procreate, is innate in all life forms. It is fundamental.

Life has even evolved two immune systems, defense systems, to keep it alive: the innate and the acquired. Both systems are made up of specialized, integral cells that prevent or eliminate diseases.

Our adaptive immune system responds to attacks by various pathogens by fighting off these attacks and then developing an immunological memory of that pathogen so that future attacks are met with a more effective immune response. This is how vaccination works.

Our innate system is so-called because it originated so far back in evolutionary times as to have become part of us. The adaptive immune system could merely be the continuation, the refining, of the innate system.

The acknowledgement of Darwin's phrase "the *struggle* for life" indicates a force exists that is violently fighting for life. All life forms are struggling to stay alive. The desperate need for life to survive, to live no matter what is mind boggling. Life is hungry for survival, but what is this "force" that drives it?

It has had many names over many years. The Greek philosopher Posidonius (c. 135 BCE – 51 BCE) described a "vital force" radiated by the sun upon all living things. Arthur Schopenhauer (1788-1860) a German philosopher, called it the "will to live." In his 1907 book *Creative Evolution*, French philosopher Henri Bergson called the force "élan vital," translated as "vital impetus" or "vital force". Other names have included vitalism, vital spark, energy, and even animism (the religious belief that all creatures, all things, possess a spiritual essence). Many of these terms connoted the idea of a "soul" giving them a religious slant which, unfortunately still exists to this day. To eliminate any religious connotations a neutral term needs to be coined, a combination of both life and force. Let's call it "liforce."

The liforce that drives speciation is potent, so potent that it moves to fill any and all available niches. It is like the need for water. It will not be denied lest it end. It may be that the liforce itself is just trying to survive, to do its due diligence, like gravity, in accordance with the laws of the universe.

Liforce has a mind of its own. Life *needs* to live and continue living. It is hungry for life and has a voracious appetite. This liforce is its own present, local, active, innate intelligence. There need not be anything special, anything outside of science to account for liforce. We only need a natural law within the complexity of the cosmos that caused life. It may be accomplishing something that the cosmos needs. It may have an aim, or a purpose—or it may not. It may be fulfilling a need—or it may not.

And it can have competence either with or without comprehension.

Read the above sentence again.

The liforce need not be aiming at an "omega point" or "noosphere," but it does seem to have an orthogenesis, to have an innate tendency to evolve in a definite direction, towards a goal, namely its own continued existence. In the movie "Jurassic Park," Dr. Ian Malcolm (played by Jeff Goldblum) says, "Life, uh, finds a way."

Life reproduces. It has succeeded, so far, and goes on continuing to reproduce. Life seems to have a memory, or, perhaps, a strategy. It certainly seems to have some sort of intelligence because as it evolves it will sometimes drop some genes with a sort of "gamble," or "knowing," that there are pathways that exist to recover what has been abandoned.

Liforce is a fortiori reality.

BUT...here's the head-shaker! The only way life can exist is by eating itself. So life fights like hell to survive at the same time it is forced to consume itself.

TELEOLOGY AND TELEONOMY

How does this liforce determine in which direction to mutate? And if it does, does not this infer teleology?

Teleology—quo vadis ("Where are you going?")—is the idea that a process has an end point, a purpose, a goal. In the case of evolution, teleology means that species are evolving toward an end, an omega point, that there may have been a cosmic purpose, with intent, that existed before the appearance of life.

Most evolutionary biologists reject this idea, hence the need for *random* mutations. There is no "mind" behind the direction of evolution. In fact, evolution has no mind, has no direction. It just is. It is functioning haphazardly; (maybe—maybe not) going nowhere in particular. It has no purpose except staying alive— the Road Runner forever avoiding the pursuit of Wile E. Coyote.

Take light for example. Modern physics classifies light as both a particle (a photon), and a wave. If a ray of light is like a string of photons, like a string of pearls, depending on from where it is viewed, it might look like a wave or a necklace.

The French lawyer/mathematician Pierre De Fermat found that when light travels from one medium into another, e.g. from air through glass, it always takes the path of least time. (How does it "know" time?) Light always chooses to minimize time over distance, De Fermat's "principle of least time".

A Dutch scientist, Willebrord Snellius, (born Willebrord Snel van Royen (1580–1626) studied the refraction of light and determined the formula for the relationship between the angles of incidence and refraction. It is called Snell's law although it was actually first discovered by the Persian scientist Ibn Sahl back in 984.

Visible light, also known as white light, is made up of all the beautiful colors of the rainbow—in order, red, orange, yellow, green, blue, indigo and violet (Roy G. Biv). Each color has a different frequency and each frequency bends at a different angle.

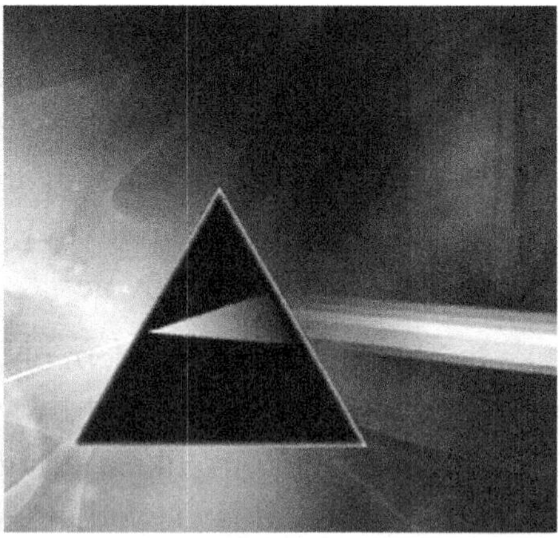

"PRISME"

Snell's Law predicts the exact angle by which light bends depending on the materials through which it is passing. The denser the material the more molecules it encounters slowing it down, changing its angle. This would seem to indicate agency, i.e. teleology. How could light know what path to choose, which path was the fastest (least time—why time over distance?) As light encounters a different medium the path is immediately obvious at the point of, and in direction of, least resistance. Light does not "know" where to go. It just follows the "laws of physics", just like gravity, just like liforce. Light selects the "fastest" route. That is nature's "law," nature's way—efficiency. This phenomenon is explained away as a fundamental property of waves which merely kicks the can down the road until we discover an explanation, a cause, more fundamental.

Teleonomy is a relatively new word, coined in 1958 by Colin Pittendrigh (1918-1996), biologist, father of chronobiology (study of circadian rhythms). His purpose was to lessen the confusion among biologists whose language continued to contain teleological phrases implying that evolution was goal oriented.

Teleonomy means that the evolutionary processes operate according to a program, that they are the products of evolution's history, its adaptation for reproductive success. The purpose of the new word was to give a separate biological meaning to the idea of Teleonomy to eliminate intention, purpose, or goal, i.e. actions without foresight.

But, from whence comes the program that life follows? It is self propagated. It is driven by life's need to exist and persist to exist, teaching itself, inch by inch, bit by bit, byte by byte, the path ahead. It is a dumb force. It has no consciousness of self. It flows on like water toward the ocean driven by that other unseen force—gravity.

Obviously, liforce does have a purpose, an end. Its purpose is to survive, procreate, and propagate in order to fulfill, perhaps, its cosmologically driven raison d'être. Natural selection works on the liforce of organisms. The two are not mutually exclusive. They are mutually embracing. There is no teleology here other than liforce—organisms following the impetus for life that drives them, a theory that begs no question. So, there is a teleological|teleonomic aspect to this understanding of evolution, but it in no way is connected to any kind of universal godhead. It is driven only by forces of nature not yet understood. We must not confuse cosmology with theology.

VARIOUS THEORIES ABOUT THE ORIGINS OF LIFE

From the moment life first formed out of some primordial soup (or RNA organic crystals), it has been on a fight to survive and procreate. We do not yet know the exact process by which life was first formed but one supposition can be seen in spontaneous combustion. Just as a pile of oily rags can burst into flames in a hot environment, so too could the right combinations of organic/inorganic elements, simmering in a pool of primordial soup under the right pressure, temperature and pH (hydrogen-ion concentration) conditions, spontaneously combust into the organic building blocks of life. Scientists have recently (2016) cooked up complex sugars (ribose) by exposing simulated space ice to low pressure and temperature (-195 C) and UV light.[42] But remember, this spontaneity is driven by natural laws, laws of physics.

The increase in the diversity of species may be the result of higher levels of oxygen, the result of photosynthesis of ocean algae and other plants converting sunlight into energy. The creation of oxygen may have resulted in the Cambrian explosion of diversity of life. Evolution takes over and, voila, here we are.

Another, more recent hypothesis of an "RNA-world" origin of life has been proposed and is, for now, the most accepted common explanation. It proposes that back in the earliest stages of evolution our last universal common ancestor (LUCA), the life form *prior* to cells, evolved from some kind of simple self-replicator. It is known that the necessary biological building blocks were present back then and that one of them—possibly a strand of RNA that could replicate RNA from an RNA template—may have been the culprit.

An alternative theory, one that also stems from the principle that evolution requires replication, is that of a nucleopeptide replicator which proposes that before life appeared on Earth there were mixtures of RNA and amino acids that were able to spontaneously form polypeptide and RNA-like chains and that these chains gave rise to our most fundamental self replicator, the one that was able to carry out functions necessary for life. This theory assumes random mutations without considering the possibility of non-random epimutations which would be following the same fundamental laws of nature.[43]

In a paper published in *Science* in October 2019, Thomas Carell, an organic chemist, and his associates add ammunition to the "RNA world" hypothesis. They proposed that about 4 billion years ago a thick crust of the organic crystals, A, U, C, and G, (adenine, uracil, cytosine and guanine) layered the Earth. Under the environmental conditions of primitive Earth (temperature, pressure, pH), combined with water and nitrogen, life began as self-replicating RNA-based genes that later developed the ability to store genetic information in DNA.

Carell goes on to suggest that there is a "strong indication" that these chemical reactions are not exceedingly lucky, that they could also happen on other planets. Could the universe be teeming with life?

Another scientist is inclined to agree. As recently as May 18, 2020, Columbia University astronomer David Kipping published in the *Proceedings of the National Academy of Sciences* that, "The case for a universe teeming with life emerges as the favored bet. The search for intelligent life in worlds beyond Earth should be, by no means, discouraged."[44]

In a quick *Reader's Digest* scenario of the earliest life forms according to NASA which has an ongoing interest in life everywhere in the universe (and beyond?), LUCA started life's journey some 4 billion years ago. It was probably born deep underground, near volcanoes or deep in the ocean near thermal vents in iron, sulfur, metal-rich environments. It did not breathe air and made its own food metabolizing these elements using hydrogen, carbon dioxide and nitrogen to turn them into organic compounds such as ammonia. Life would take non-organic carbon and turn it into organic carbon without the presence of light. It was chemical energy that drove these early life forms. Gradually, they laid down mats upon which other life forms could feed. The earliest fossil evidence we have of these layers of sentiment deposited by microbes are stromatolites dating back to 3.7 billion years ago. [45] Later, photosynthetic bacteria evolved using sunlight to build sugars from carbon dioxide and water, or so the story goes.

These hypotheses are of course just that, theories, suppositions, and conjectures just as are the *genes first* (a pre-RNA world) or *metabolism first* ideas. We are still searching for answers.

METABOLISM – THE OUROBOROS PROBLEM

As previously mentioned, one of the basic properties of life is metabolism, the conversion of food/fuel to energy.

But where do organisms get food/fuel/energy?

We eat one another.

Big fish eats little fish eats littler fish.

This produces quite a paradox. If we need to eat life for life to continue to exist and flourish as it does, in fact flourish in ever increasing numbers, are we not killing ourselves even as we struggle to survive?

There is an ancient Egyptian symbol depicting a serpent or dragon eating its own tail. It's first found in a funereal book in the tomb of Tutankhamen, 14th century BC. It is called an ouroboros literally meaning tail-devourer in Greek. It is often interpreted as the cycle of life, death, rebirth, the eternal cycle of renewal, cyclical time.

Ouroboros
Alchemy: The Sacred Art
Courtesy of S. Klossowski de Rola, Alchemie

We eat life to sustain life. Just as a snake sheds its skin in order to grow, the ouroboros models how we can be reborn again and again. It also illustrates the paradox of life eating life in order to survive. The cycle of life, despite being at odds with itself, exists in union with an incessant, ongoing struggle.

In an attempt to find why life exists at all, why we are eating each other in order to exist, an examination of its place in the cosmos seems necessary. Does life have a purpose? Are we eating each other in order to reduce compounds to elements? Is life trying to hasten entropy?

Could we be viewing entropy upside down? Could nothingness, stasis, be the goal of the universe using life to achieve its end? Are the natural laws of the universe using the natural law of the liforce for their own purposes? Alas, another conundrum. Will we, can we, ever know?

OUR GENETIC CODE

Whatever the origin of life, another enigma is the origin of our genetic code. It is nearly universal, and highly specific, i.e. non-random. There are several theories, none of which hold much sway in today's thinking, and all are compatible with a "frozen accident" theory.

The "frozen accident" notion proposes that our standard genetic code has no special properties but was fixed accidently simply because all extant life forms share the same common ancestor.[46] The frozen accident concept is similar to that of the spontaneous combustion theory for the beginnings of life—nothing more than the "accidental" occurrence of the right elements at the right time under the right circumstances. Think Occam's Razor (a logic principle to keep things as simple as possible),. Think the random mutations theory. These ideas are non-answers—are merely interesting guesses.

NEURAL NETWORKS

But, again, how does the liforce "decide" in which direction to mutate? Remember teleology?

It does not decide. It merely continues on its quest to survive, determined by the successes it has already had throughout time, by fine tuning the cybernetic application of both positive and negative feedback. After either the spontaneous creation or the RNA world view—whatever—driven by the liforce, epimutations evolved driven by the environment. Some mutations survived by adapting. That is a plus. Some failed. That is a minus. By adding up the pluses and subtracting the minuses, life evolved in a direction that led directly here-now. Life follows its genetic programming. Teleonomy connotes purpose to the *program*, i.e. a force behind liforce, a law of the cosmos we are struggling to understand—the purpose of the program and not the organism. This is a problem we are still trying to solve.

It must have been dicey for those first few organisms teetering on the seesaw of survival—one wrong move and, whoops, no life at all, extinction. But the liforce is so strong it will simply not take no for an answer, perhaps like that apple *had* to fall into Newton's lap (or on his head).

In 1950, Claude Shannon (1916-2001), known as the father of information theory created an electromagnetic mouse he called "Theseus." This mouse was able to negotiate a variable maze to find a target. Having traversed the maze, the mouse could then be placed anywhere in the maze and, by virtue of its previous experience, find the target. If placed in a new location, it would search until it found a familiar spot from whence it would proceed to the target adding this new knowledge to its memory.[47]

This is precisely how scientists developing artificial intelligence (AI) are proceeding.

In 2017, AI researchers, after years of brute force computations, developed a "*genetic* [emphasis mine] algorithm to evolve a network with the goal of achieving optimal hyperparameters [hyperparameters are characteristics that define particular systems used to control learning processes] in a fraction of the time of a brute force search."[48]

But what is a genetic algorithm? "...a genetic algorithm is a metaheuristic [a higher-level procedure or design] used to generate high-quality solutions to optimization and search problems by relying on *bio-inspired* [emphasis mine] operators such as mutation, crossover and selection."[49]

And here's how AI works:

1) "Creates a population of (randomly generated) members

2) Scores each member of the population based on some goal. This score is called a fitness function.

3) Selects and *breeds* the best members of the population to produce more like them

4) Mutates some members randomly to attempt to find even better candidates

5) Kills off the rest (survival of the fittest ...)

6) Repeats from step 2. Each iteration through these steps is called a generation.

Repeat this process enough times and you [will] be left with the very best possible members of a population. Sounds like a lot evolution, right? Same deal."[50]

Could this be how evolution works? The first life forms existed in what must certainly been hostile environments. Some survived. Some did not. Those that made it established a vestigial memory of what worked and continued epimutating, continued breeding gradually becoming more and more varied.

Artificial intelligence *breeding* is where [researchers] take two members of a population and generate one or more child, where that child represents a combination of its parents.

"In our neural network case, each child is a combination of a random assortment of parameters from its parents. For instance, one child might have the same number of layers as its mother and the rest of its parameters from its father. A second child of the same parents may have the opposite. You can see how this mirrors real-world biology and how it can lead to an optimized network quickly."[51]

Very quickly indeed! The author of this study reports that as the parameter complexity increases, the genetic algorithm provides an exponential speed benefit. For the problem he addressed it cut the time for a solution to the problem submitted by a factor of nine.

So, we can see here how evolution may have proceeded, by trial and error, building up biological information over time, branching out and creating new species.

SPACE: THE FINAL FRONTIER

Man's unending curiosity has led him to every corner of planet Earth no matter how remote. It has taken humans to the moon and there are plans for man to go to Mars. Such a trip will expose the astronauts to a hostile environment unlike anything ever before experienced. A trip to Mars will probably take about 30 months including time on the ground on the planet and will expose them to isolation, confinement, microgravity, radiation, and noise on an unprecedented scale.

Scientists are already studying these problems. In 2015, NASA selected 50 year old identical twins, Scott and Mark Kelly, for their experiments. They sent Scott to the International Space Station where he floated in microgravity for 340 days while his brother remained on Earth. Neither astronaut had been in space for nearly four years before this experiment.

During the course of the study Scott periodically drew blood, urine and feces samples while doctors did the same for his twin on Earth so that comparisons between the two could be studied in depth.

Upon Scott Kelly's return to Earth, scientists found significant differences in his physiology from that of his twin brother. His immune system was in a hyperactive state which alerted them that his genes were trying to understand

and adjust to his new environment. Fluids caused his head and upper body to swell. The shape of his eyeballs changed including a thickening of his retina. The most dramatic were the changes to his genes. More than 1000 of his genes underwent epigenetic changes including the lengthening of his telomeres. Telemeres are the caps, protective shields, on the ends of each chromosome. They shorten as we age and/or experience stress, pollution, radiation, etc. When they shrink to a certain size cells stop dividing. As such they are considered a biomarker of accelerated aging. They are essential in protecting DNA ends from degradation and damage.[52]

After Scott's return to Earth, most (91.3%) of his gene-activity levels reverted to normal, but he showed a continued decrease in cognitive speed, accuracy and memory for a full six months. His immune system remained on high alert. Further results will be forthcoming as the twins grow older and the experiment continues.

FOOD FOR THOUGHT: "Time dilation" (a well confirmed fact) happens when one clock accelerates relative to a stationary clock. The clock in motion seems to run slower than the stationary clock, slower and slower the closer to the velocity of light it gets. This is not a feature of the mechanics of the clock itself. It is actually time itself that gets "dilated."

So, given our twins, Scott who moved at a higher relative speed with respect to Mark on Earth, aged less. He is younger now than Mark. Shortened telomeres anyone?[53]

INTELLIGENT DESIGN (NOT)

This disputation is not to be confused with a defense of "intelligent design" which seeks to tie cosmology to theology. There is no reason to do so. Just as the apple fell into Newton's lap (or on his head) and he discerned a teleological force he could not see, namely gravity, and just as Einstein predicted the existence of gravity waves 250 years later, long before we had the ability to detect them, we can attribute liforce to a cosmological phenomenon the purpose of which we cannot, as yet, determine.

Why is there such a thing as "life?" Is there a reason for liforce? Why does life need to consume life in order to survive? Liforce is just another form of energy. Every time we take a breath we prove it. Hopefully, liforce may someday be as equally accepted as gravity.

Darwin maintained that "There seems to be no more design in the variability of organic beings and in the action of natural selection, than in the course in which the wind blows."[54] Bob Dylan taught us that you don't have to be a weatherman to know which way the wind blows. We now know that the wind blows in a certain direction dictated by weather patterns unfolding moment by moment all around the globe. So, yes, the reason for the direction of the wind is, ultimately, discernible. The "butterfly effect" anyone?

In chaos theory the butterfly effect describes how a small change in one system can result in major changes in another. "The fluttering of a butterfly's wings can effect climate changes on the other side of the planet," says Nobel Prize winner Paul Erlich

A universal intelligence does not refer to a god with some intelligent design he/she/it has launched. Unfortunately, given the tenor of these times, an offense against intelligent design seems required.

Just as gravity is not any sort of god but, rather, a law of nature, so too is liforce an act of essentiality. Both follow scientific principles. These principles we deem "intelligent" because they are smarter than we are. They demand intelligence in our attempts to understand them, and we are doing a pretty good job of doing just that but we have a long way to go.

In Greek mythology Orthrus was a two headed dog that defended the giant Geryon's herd of red cattle. Like Orthrus the battle between the creationist's head and the atheist's head continues unabated, each rearing its head in argument. The creationist's head looks to the intelligent design of the cosmos and everything in it as proof of the existence of a capital "G" God. The atheist's head denies any intelligence behind designs of any kind. Evolution is the result of random changes made by a "blind watchmaker"—to use Richard Dawkins' oft-quoted phrase. Dawkins is a British, evolutionary biologist, atheist, and author famous for developing the concept of "memes," an entity like a gene but one that replicates ideas instead of biological information.

Why an agnostic argument is seldom proposed in these discussions is perplexing. Discussions of the randomness of mutations using neither faith nor disbelief would seem a fruitful endeavor.

In our example, both of Orthrus' heads are deaf, deaf to explanations not in concert with the pre-ordained beliefs of each side. There may or may not be a god. We will never know. That does not discount "intelligence" behind whatever is driving the cosmos as the atheists profess in their conceit. Likewise, professing a belief in something that can never be proven is the creationists' faith and their folly.

Screams of the righteousness of each position are so loud that deafness has resulted in both of Orthrus's heads as each defends its own special pleading of a double standard, criticizing the other for failing to prove its point while exempting its own. One side professes randomness while the other touts cosmic intelligence, each defending its stance as the only acceptable hypothesis.

Each side poisons the well of the other's logic, using straw-man arguments and red herrings. Straw-man attacks misrepresent their opponent's position. Red herrings introduce irrelevant suppositions to deflect attention from the real issue being debated—the randomness of mutations versus intelligent design.

With random mutations there does not have to be any explanation of evolution. Evolution just happens by chance. However, if there is a liforce there is a direction, a teleology of sorts. But that just cannot be because there is no intelligence of any kind at work here. Just ask the atheists.

But God is the intelligence behind the laws of nature. Just ask the creationists.

Perhaps David Hume, the 18th century philosopher said it best when discussing the intelligent design of the universe: "[I]ntelligent design does not violate any known facts or logical principles. So why does it meet with such fierce opposition? True, many people leap immediately from the notion of intelligent design to the theology of their particular sect, but the proper response by anyone who claims intellectual rigor is to show the hidden assumptions in *that leap of reasoning*." [Emphasis mine] So, can we now dispense with any more dialogues about theology when discussing the intelligence of the universe?

INTELLIGENCE

Sadly, even the term intelligence needs defining lest it be appropriated into this argument on either—or both—sides. Intelligence is commonly defined as the ability to acquire and apply knowledge, the ability to reason, the ability to think abstractly, mental acuteness, shrewdness.

Are humans the only example of intelligence in the universe? Are we the only beings able to suss out $e=mc^2$? Is it us alone who establish so called "scientific" rules that really do not exist except in our own minds, for our own self-satisfying amusement. We describe laws of motion, name stray particles (quarks, hadrons, bosons) that we invent in our weak attempts at understanding that which really has or may have no rules, and no laws? Is it merely chaos that we are attempting to classify for the benefit of our own egos, pronouncing axioms that are truly meaningless?

Ray Kurzweil, director of engineering at Google, author of the book, *The Singularity is Near*, asks us if perhaps the intelligence of the universe is "according to common wisdom... irrelevant to cosmological thinking. It is just a bit of froth dancing in and out of the crevices of the universe, and has no effect on our ultimate cosmological destiny." An interesting idea, he muses, but one ***not*** his own. Instead, Kurzweil prefers the view that "the universe has been set up exquisitely enough to have intelligence."

So, is there an actual system to the cosmos that we are trying to understand? If our intelligence strives for these understandings are we not searching in an ether of intelligence far beyond the capabilities of the greatest minds of all time?

Assemble a group of these great minds: Einstein, Marie Curie, Confucius, Newton, Hepatica of Alexandria, Haile Selassie, George Carlin, Muhammad ibu Zakariya al-Razi, Hannah Arendt, Maimonides, Plato and throw in Richard Dawkins to make it an even dozen. All of these great minds, on their best days, without distractions, all mind-melding together would be unable to explain the intelligence of the universe while hearing a congress of baboons in the background laughing uproariously. In striving to grok this intelligence, we should acknowledge that it is beyond our knowing. We should bow our heads and weep before its magnificence.

MATH SIMPLIFIED

Although science keeps adjusting basic estimates (which also vary depending on sources), according to present scientific theories, the Big Bang occurred about 13.8 billion years ago. Earth is about 4.5 billion years old and the first life forms originated approximately 3.7 billion years ago. Homo sapiens evolved around 300,000 years ago, but culture, e.g. art and complex tools, only evolved 65,000 to 50,000 years ago.

Whoa! Too much math? Billions, millions, tens of thousands of years are concepts difficult to wrap our heads around.

One way of understanding the expanse of evolutionary time is to compare it to something with which we are familiar. Let's use the length of a football field, 100 yards, as our benchmark. Humans, *Homo sapiens*, first appeared about 300,000 years ago. Calculating that time scale on our football field analogy, starting with the big bang at one end zone man would show up less than one quarter of an inch from the far end zone.

AN EVOLUTION REFRAIN

The centuries' ongoing debate over nature vs. nurture—which one determines who we are—appears to have been answered. It's both.

The term "nature vs. nature" was actually coined by Darwin's cousin, Francis Galton around 1875. Nature is what an individual inherits from its parents, its genes and physical structure. Nurture is the environment into which that nature is born and lives. The two then work both in unison and opposition until death, fine tuning each organism on its journey into the future.

Although scientists bicker about details in this debate, there are some basic characteristics upon which they generally agree. There are a number of theories of how life began that are held in common, and, regardless of the theory presented, once life was established evolution took over.

A population of self-replicating organisms sat simmering in pots of primordial soup, each within differing environments, some too hot, some too cold and some just right. Those in the pot too hot were stressed in one direction. Those in the pot too cold were stressed in another while those in the Goldilocks pot hummed along quite happily.

These stressed organisms, each in its own private environment, developed epigenetic markers which enabled or suppressed gene expression, all absent randomness, all driven by the liforce. As these organisms procreated they passed these markers onto their progeny. Given that these stresses continued with enough intensity over sufficiently-long evolutionary time periods, the epimutations eventually became fixed into the lineage by altering the DNA itself. Voila—the origin of a new species.

Survival of the fittest also includes cultural dynamics that promote survival. A new understanding of nature's laws, a new tool, a new word, a new idea, and, yes, a new symphonic composition may increase the propensity of life to survive and procreate. Culture counts! It *is* the environment.

Earth has experienced five mass extinction events due to natural causes as well as "normal" extinctions that occur naturally and periodically.

Scientists describe mass extinctions events as those in which 75% of existing species die off over a period of geological time of two million years.

These events are described by time periods in which they occurred. The first is identified as the End Ordovician, 444 million years ago (mya). This particular event was caused by an uplift of tectonic plates.

The latest, called the End Cretaceous, occurred 65 mya and was caused by an asteroid strike in Yucatán, Mexico. This was the event that killed off the dinosaurs. Others were caused by volcanic eruptions and, believe it or not, rapid plant growth which resulted in severe global cooling.

Sadly, we may now be on the verge of a sixth event, this one caused by human stupidity, human complacency, human selfishness, human wars, and/or over-population. It is estimated that there are presently about 8 million species here on Earth, but somewhere between 90-99 percent of all species that ever existed have already gone extinct. Meanwhile, the liforce is forever potent. It is moving inexorably forward—until the sun blinks out.

Who/What will survive?

EVOLUTION'S SYMPHONY

So, we have come full circle. Grandma made you fat and it's not your fault. Blame the environment but forego that second helping of Ben & Jerry's Phish Food ice cream doused with chocolate syrup, and whipped cream topped with a cherry. We don't want to expand our cycle of obesity. We are fat enough already.

Back to our symphony.

A liforce erupts from the dissonant drama of chaos beginning evolution's symphony as interpreted by its conductor, the environment. A collision of cymbals ignites the spontaneous combustion of conception, piano ivories tinkle with life, drums pulse rhythm, hearts beat ones and zeros. A single violin string hums a note quickly picked up by a class of cellos as new life forms mate in harmony.

Listen!

A clarinet announces an overture of alleles vibrating melodies. French horns pronounce ever expanding chords of speciation; brass blast blows of violent struggles, bassoons the anguish of extinction.

Can you hear that sax undulating sex? A flute trills hope for survival of the most fit; trombones slip-slide between epimutations; an oboe haunts, penetrating searches for the liforce; as an organ suddenly announces self-aware intelligence in reverberating resonances.

The symphony plays its unwritten score of circling celestial spheres—while, in the background, a vexing, curious static hisses its accompaniment.

Exult yet weep more. Weep in the confused desperation of never knowing, never fully understanding. Then, wipe your eyes, one after the other, and look to the horizon. One day maybe...one day...just maybe...

ADMONITION

I end this disputation with the admonition that these are only my own personal reflections. There are much-more-learned women and men than I who bring potent, conflicting arguments to this table. I look forward to learning from them. Curiosity is that vexing static, hissing forever in the back ground, continuing its searches for original causes.

SCIENCE MARCHES ON

Even as I type this science marches on. Both new and ongoing discoveries, and theories onrushing into the future (some dramatic, some mundane) add to the bank of knowledge moment by moment. Since published in 2020 new research has expanded the thesis of this work and will continue to do so. Science, like time, marches on.

CONSOLATION

I leave you with a poem about Charles Darwin by the Nobel Prize winning Polish poetess Wislawa Szymborska

CONSOLATION

They say he read novels to relax,
But only certain kinds:
nothing that ended unhappily.
If anything like that turned up,
enraged, he flung the book into the fire.

True or not,
I'm ready to believe it.

Scanning in his mind so many times and places,
he'd had enough of dying species,
the triumphs of the strong over the weak,
the endless struggles to survive,
all doomed sooner or later.
He'd earned the right to happy endings,
at least in fiction
with its diminutions.

Hence the indispensable
silver lining,
the lovers reunited, the families reconciled,
the doubts dispelled, fidelity rewarded,
fortunes regained, treasures uncovered,
stiff-necked neighbors mending their ways,
good names restored, greed daunted,
old maids married off to worthy parsons,
troublemakers banished to other hemispheres,
forgers of documents tossed down the stairs,
seducers scurrying to the altar,
orphans sheltered, widows comforted,
pride humbled, wounds healed over,

prodigal sons summoned home,
cups of sorrow thrown into the ocean,
hankies drenched with tears of reconciliation,
general merriment and celebration,
and the dog Fido,
gone astray in the first chapter,
turns up barking gladly
in the last.

Thank you Harper Collins.
"Consolation" from MAP: Collected and Lost Poems by Wislawa Szymborksa, translated from the Polish by Stanislaw Baranczak and Clare Cavanagh.

END NOTES

1. [1] https://jamesclear.com/wp-content/uploads/2015/01/tessa-roseboom-effects-of-prenatal-exposure-to-the-dutch-famine-on-adult-disease-in-later-life.pdf

2. [2]Carey, Nessa.*The Epigenetics Revolution: How Modern Biology is Rewriting Our Understanding of Genetics, Disease, and Inheretance.* Columbia University Press. 2013

3. [3] Ibid

4. [4] https://www.nytimes.com/2018/01/31/science/dutch-famine-genes.html

5. [5] https://obgyn.onlinelibrary.wiley.com/doi/epdf/10.1111/j.1471-0528.2008.01822.x

6. [6] Carey, Nessa.*The Epigenetics Revolution: How Modern Biology is Rewriting Our Understanding of Genetics, Disease, and Inheretance*, p2, Columbia University Press. 2013

7. [7] https://www.sciencedaily.com/releases/2014/09/140929105358.htm

8. [8] Joblonka, Eva and Lamb, Marion J., *Evolution in Four Dimensions*, p.142, MIT Press, 2005

9. [9] Jablonka, p138

BIBLIOGRAPHY

A critical view on transgenerational epigenetic inheritance in humans https://www.nature.com/articles/s41467-018-05445-5

A Selective Sweep on a Deleterious Mutation in ***CPT1A*** in Arctic Populations https://www.sciencedirect.com/science/article/pii/S0002929714004224

Andean, Tibetan, and Ethiopian patterns of adaptation to high-altitude hypoxia https://academic.oup.com/icb/article/46/1/18/661204

Characteristics of Life - What is Life? http://www.bellaonline.com/articles/art18278.asp

Dilation of Time: https://www.physlink.com/Education/Askexperts/ae260.cfm

Ennis, Cath & Pugh, Oliver. *Epigenetics – A Graphic Guide*, Icon Books, 2017.

Environmental stress and epigenetic transgenerational inheritance https://www.ncbi.nlm.nih.gov/pmc/articles/PMC4244059/

Environmentally induced epigenetic transgenerational inheritance of sperm epimutations promote genetic mutations https://www.ncbi.nlm.nih.gov/pmc/articles/PMC4622673/

Epigenetic Inheritance and Its Role in Evolutionary Biology: Re-Evaluation and New Perspectives https://www.ncbi.nlm.nih.gov/pmc/articles/PMC4929538/

Ernst Mayr and the modern concept of species https://www.pnas.org/content/102/suppl_1/6600

For an in depth examination of speciation in asexual organisms see *Lost Sex The Evolutionary Biology of Parthenogenesis* Edited by Peter Van Dijk, Koen Martens, Isa Schön.

Genetic Drift https://evolution.berkeley.edu/evolibrary/article/evo_24

High-altitude living has changed more than just the genes of some Peruvians

High-altitude living has changed more than just the genes of some Peruvians | Science | AAAS (sciencemag.org)

How the Flu Virus Can Change: "Drift" and "Shift" https://www.cdc.gov/flu/about/viruses/change.htm

Mutations can reveal how the coronavirus moves—but they're easy to over interpret

https://www.sciencemag.org/news/2020/03/mutations-can-reveal-how-coronavirus-moves-they-re-easy-overinterpret?utm_campaign=news_weekly_2020-03-13&et_rid=475950291&et_cid=3244543

On the Dual Nature of Chance in Evolutionary Biology and Paleobiology

https://www.researchgate.net/publication/23740539_On_the_Dual_Nature_of_Chance_in_Evolutionary_Biology_and_Paleobiology

Origin and evolution of the genetic code: the universal enigma https://www.ncbi.nlm.nih.gov/pmc/articles/PMC3293468/

Persistent epigenetic differences associated with prenatal exposure to famine in humans

https://www.ncbi.nlm.nih.gov/pmc/articles/PMC2579375/

Phenotypic Plasticity https://advances.sciencemag.org/content/6/21/eaba3388?utm_campaign=toc_advances_2020-05-22&et_rid=475950291&et_cid=3336736

Redon R, Ishikawa S, Fitch KR, Feuk L, Perry GH, Andrews TD, et al. (November 2006). "Global variation in copy number in the human genome". *Nature*, 444 (7118): 444–54. Bibcode:2006Natur.444..444R. doi:10.1038/nature05329. PMC 2669898. PMID 17122850.

Schmitz RJ, Schultz MD, Urich MA, Nery JR, Pelizzola M, Libiger O, Alix A, McCosh RB, Chen H, Schork NJ, Ecker JR. Patterns of population epigenomic diversity. Nature. 2013;495:193–198. [PMC free article] [PubMed] [Google Scholar]

Skinner MK, Manikkam M, Guerrero-Bosagna C. Epigenetic transgenerational actions of environmental factors in disease etiology. Trends Endocrinol Metab. 2010;21:214–222. [PMC free article] [PubMed] [Google Scholar]

Speciation: https://www.nationalgeographic.org/encyclopedia/speciation/

Study documents paternal transmission of epigenetic memory via sperm

https://phys.org/news/2018-10-documents-paternal-transmission-epigenetic-memory.html

Sugars cooked up from 'comet ice' https://www.bbc.com/news/science-environment-35991312

The Effects of Deleterious Mutations on Evolution at Linked Sites https://www.genetics.org/content/190/1/5

The road to modularity, Wagner GP[1], Pavlicev M, Cheverud JM.
https://www.ncbi.nlm.nih.gov/pubmed/18007649

Transgenerational Epigenetic Inheritance: myths and mechanisms
https://www.ncbi.nlm.nih.gov/pmc/articles/PMC4020004/

Transgenerational epigenetic inheritance: prevalence, mechanisms, and implications for the study of heredity and evolution
https://www.ncbi.nlm.nih.gov/pubmed/19606595

Waddington's Assimilation, A Fact or A Mere Philosophy Shaped in The Lamarckian Mold: A Genomic Inquiry

http://www.enlivenarchive.org/articles/waddingtons-assimilation-a-fact-or-a-mere-philosophy-shaped-in-the-lamarckian-mold-a-genomic-inquiry.html#1

What is Darwin's Theory of Evolution? https://www.livescience.com/474-controversy-evolution-works.html

https://www.gesundheitsindustrie-bw.de/en/article/news/endosymbiosis-and-horizontal-gene-transfer

Bonduriansky, Russell and Day, Troy, *Extended Heredity*, pp. 10/11, Princeton University Press, 2018

Carey, Nessa. *The Epigenetics Revolution: How Modern Biology is Rewriting Our Understanding of Genetics, Disease, and Inheritance*, Columbia University Press. 2013.

Charles Darwin, *The Life and Letters of Charles Darwin*, ed. Francis Darwin, vol. 1 (New York: Appleton, 1887), 278–279

Francis, Richard C. *How Environment Shapes Our Genes*, W. W. Norton, 2012.

http://www.eebweb.arizona.edu/faculty/birky/Readings/Birky07AsexualSpecies.pdf]

https://academic.oup.com/icb/article/46/1/18/661204https://hypebeast.com/2012/4/town-of-runners-a-documentary-about-incredible-ethiopian-distance-runners

https://advances.sciencemag.org/content/6/21/eaba3388?utm_campaign=toc_advances_2020-05-22&et_rid=475950291&et_cid=3336736

https://astrobiology.nasa.gov/news/looking-for-luca-the-last-universal-common-ancestor/

https://blog.coast.ai/lets-evolve-a-neural-network-with-a-genetic-algorithm-code-included-8809bece164

https://en.wikipedia.org/wiki/Copy-number_variation

https://en.wikipedia.org/wiki/Genetic_algorithm

https://en.wikipedia.org/wiki/Genetic_drift

https://hypebeast.com/2012/4/town-of-runners-a-documentary-about-incredible-ethiopian-distance-runners

https://jamesclear.com/wp-content/uploads/2015/01/tessa-roseboom-effects-of-prenatal-exposure-to-the-dutch-famine-on-adult-disease-in-later-life.pdf

https://metro.co.uk/2018/04/19/tribe-of-fish-people-evolved-huge-spleens-so-they-can-dive-to-the-bottom-of-the-ocean-7481734/

https://obgyn.onlinelibrary.wiley.com/doi/epdf/10.1111/j.1471-0528.2008.01822.x

https://phys.org/news/2019-10-scientists-recreate-flies-mutations-monarch.html

https://phys.org/news/2020-10-yeast-yields-insights-longstanding-evolution.html

https://www.bbc.com/news/science-environment-35991312

https://www.cell.com/cell/fulltext/S0092-8674(11)01341-

9?_returnURL=https%3A%2F%2Flinkinghub.elsevier.com%2Fretrieve%2Fpii%2FS0092867411013419%3Fshowall%3Dtrue

https://www.cell.com/cell/fulltext/S0092-8674(18)30386-6

https://www.cell.com/trends/ecology-evolution/fulltext/S0169-5347(20)30003-3?dgcid=raven_jbs_aip_email

https://www.genome.gov/Genetic-Disorders/Sickle-Cell-Disease

https://www.medicaldaily.com/bacteria-must-compete-resources-survive-even-if-it-means-fight-death-234701

https://www.ncbi.nlm.nih.gov/pmc/articles/PMC3293468/

https://www.ncbi.nlm.nih.gov/pmc/articles/PMC4622673/

https://www.ncbi.nlm.nih.gov/pmc/articles/PMC4804638/

https://www.news-medical.net/news/20170427/Researchers-discover-new-mechanism-of-epigenetic-inheritance-in-fruit-flies.aspx

https://www.nytimes.com/2018/01/31/science/dutch-famine-genes.html

https://www.pbs.org/wgbh/evolution/educators/course/session2/explain_c_pop2.html

https://www.physlink.com/Education/Askexperts/ae260.cfm

https://www.pnas.org/content/early/2020/05/12/1921655117

https://www.quantamagazine.org/why-sex-biologists-find-new-explanations-20200423/

https://www.sciencealert.com/scientists-just-captured-the-actual-flash-of-light-that-sparks-when-sperm-meets-an-egg

https://www.sciencealert.com/scientists-observe-epigenetic-memories-passed-down-for-14-generations-most-animal#.WPvetdCGJKw.facebook

https://www.sciencedaily.com/releases/2014/09/140929105358.htm

https://www.sciencedirect.com/science/article/pii/S0002929714004224

https://www.sciencemag.org/news/2014/07/tibetans-inherited-high-altitude-gene-ancient-human

https://www.sciencemag.org/news/2019/08/genetic-mutation-makes-people-feel-rested-just-6-hours-sleep

https://www.sciencemag.org/news/2020/07/ancient-microbial-arms-race-sharpened-our-immune-system-also-left-us-vulnerable?utm_campaign=news_weekly_2020-07-31&et_rid=475950291&et_cid=3430879

https://www.sciencenewsforstudents.org/article/how-year-space-affected-scott-kellys-health

Jablonks, Eva and Lamb, Marion J., *Evolution in Four Dimensions*, MIT Press, 2005

Monod, Jacques, *Chance and Necessity*, p. 101-102 English translation by William Collins Sons and Co. Ltd, Glasgow, 1972

Proceedings of the Royal Society B: Biological Sciences. "Terrestrial habitats decouple the relationship between species and subspecies diversification in mammals" by Laura van Holstein and Robert A. Foley, 17 March 2020

Theseus | History of AI

Thomas McGann is a native New Yorker, a graduate of the United States Coast Guard Academy. Besides a military career that included a tour in Vietnam, Tom formed Seafarers Investing Company along with two fellow Academy classmates. They ran restaurants, retail outlets, marinas, a construction company and several real estate ventures.

Tom spent ten years digging clams and was a house painter for another ten. Other jobs included bartending, waiting tables and driving trucks.

But Tom's passion is writing. He is the author of the young adult fantasy, *The Riddle of Riddles*, the gritty motorcycle novel *Chance*, non-fiction *Historical Anecdotes from the Fire Island News*, the scientific treatise *Grandma Made You Fat – It's Not Your Fault*, numerous short stories, essays, plays and poetry.

Sailing, motorcycling and reading are his hobbies but family is his life.

---

10. [10] https://www.news-

medical.net/news/20170427/Researchers-discover-new-mechanism-of-epigenetic-inheritance-in-fruit-flies.aspx

11. [11] https://phys.org/news/2019-10-scientists-recreate-flies-mutations-monarch.html

12. [12] https://www.sciencealert.com/scientists-observe-epigenetic-memories-passed-down-for-14-generations-most-animal#.WPvetdCGJKw.facebook

13. [13] https://www.cell.com/cell/fulltext/S0092-8674(11)01341-9?_returnURL=https%3A%2F%2Flinkinghub.elsevier.com%2Fretrieve%2Fpii%2FS0092867411013419%3Fshowall%3Dtrue

14. [14] Francis, Richard C. *How Environment Shapes Our Genes,* pp. 114,115, W.W. Norton, 2012

15. [15] Jablonka, Eva and Lamb, Marion J., *Evolution in Four Dimensions*, pp 97,98,322, MIT Press, 2005

16. [16] https://www.pbs.org/wgbh/evolution/educators/course/session2/explain_c_pop2.html

17. [17] Francis, Richard C. *How Environment Shapes Our Genes*, p. 46, 47. W. W. Norton, 2012.

18. [18] ***Proceedings of the Royal Society B: Biological Sciences***. "Terrestrial habitats decouple the relationship between species and subspecies diversification in mammals" by Laura van Holstein and Robert A. Foley, 17 March 2020,

19. [19] https://www.sciencemag.org/news/2014/07/tibetans-inherited-high-altitude-gene-ancient-human

20. [20]https://academic.oup.com/icb/article/46/1/18/661204https://hypebeast.com/2012/4/town-of-runners-a-documentary-about-incredible-ethiopian-distance-runners

21. [21] https://hypebeast.com/2012/4/town-of-runners-a-documentary-about-incredible-ethiopian-distance-runners

22. [22] https://www.cell.com/cell/fulltext/S0092-8674(18)30386-6

23. [23] https://metro.co.uk/2018/04/19/tribe-of-fish-people-evolved-huge-spleens-so-they-can-dive-to-the-bottom-of-the-ocean-7481734/

24. [24] https://www.sciencedirect.com/science/article/pii/S0002929714004224

25. [25] https://www.sciencemag.org/news/2019/08/genetic-mutation-makes-people-feel-rested-just-6-hours-sleep

26. [26] https://www.genome.gov/Genetic-Disorders/Sickle-Cell-Disease

27. [27] https://www.quantamagazine.org/why-sex-biologists-find-new-explanations-20200423/

28. [28] http://www.eebweb.arizona.edu/faculty/birky/Readings/Birky07AsexualSpecies.pdf]

29. [29] https://en.wikipedia.org/wiki/Genetic_drift

30. [30] https://www.ncbi.nlm.nih.gov/pmc/articles/PMC4804638/

31. [31] https://www.sciencemag.org/news/2020/07/ancient-microbial-arms-race-sharpened-our-immune-system-also-left-us-vulnerable?utm_campaign=news_weekly_2020-07-31&et_rid=475950291&et_cid=3430879

32. [32]https://advances.sciencemag.org/content/6/21/eaba3388?utm_campaign=toc_advances_2020-05-22&et_rid=475950291&et_cid=3336736

33. [33] Bonduriansky, Russell and Day, Troy, *Extended Heredity*, pp10/11, Princeton University Press,

2018

34. [34] https://www.tandfonline.com/doi/full/10.1080/15592294.2015.1062207

35. [35] https://en.wikipedia.org/wiki/Copy-number_variation [Copy number variation is a phenomenon in which sections of the genome are repeated and the number of repeats in the genome varies between individuals in the human population... there had been discussion connecting copy number variations to gene families. Gene families are defined as a set of related genes that serve similar functions but have minor temporal or spatial differences and these genes likely derived from one ancestral gene The main reason copy number variations are connected to gene families is that there is a possibility that genes in a family may have derived from one ancestral gene which got duplicated into different copies. Mutations accumulate through time in the genes and with natural selection acting on the genes, some mutations lead to environmental advantages allowing those genes to be inherited and eventually clear gene families are separated out.]

36. [36] https://www.ncbi.nlm.nih.gov/pmc/articles/PMC4622673/ Environmentally induced epigenetic transgenerational inheritance of sperm epimutations promote genetic mutations (nih.gov)

37. [36] Ibid

38. [38] Endocrine Disruptor Induction of Epigenetic

Transgenerational Inheritance of Disease (nih.gov)

39. [39] https://www.sciencealert.com/scientists-just-captured-the-actual-flash-of-light-that-sparks-when-sperm-meets-an-egg

40. [40] Monod, Jacques, *Chance and Necessity*, p. 101-102 English translation by William Collins Sons and Co. Ltd, Glasgow, 1972. In this work, by "epigenesis" Monod specifies that it is "not in relation to any theory, but in reference to all processes of structural and functional development."

41. [41] https://www.medicaldaily.com/bacteria-must-compete-resources-survive-even-if-it-means-fight-death-234701

42. [42] https://www.bbc.com/news/science-environment-35991312

43. [43] https://www.cell.com/trends/ecology-evolution/fulltext/S0169-5347(20)30003-3?dgcid=raven_jbs_aip_email

44. [44] https://www.pnas.org/content/early/2020/05/12/1921655117

45. [45] https://astrobiology.nasa.gov/news/looking-for-luca-the-last-universal-common-ancestor/

46. [46] https://www.ncbi.nlm.nih.gov/pmc/articles/PMC3293468/

47. [47] Theseus | History of AI

48. [48] https://blog.coast.ai/lets-evolve-a-neural-network-with-a-genetic-algorithm-code-included-8809bece164

49. [49] https://en.wikipedia.org/wiki/Genetic_algorithm

50. [50] https://blog.coast.ai/lets-evolve-a-neural-network-with-a-genetic-algorithm-code-included-8809bece164

51. [51] Ibid

52. [52] https://www.sciencenewsforstudents.org/article/how-year-space-affected-scott-kellys-health

53. [53] https://www.physlink.com/Education/Askexperts/ae260.cfm

54. [54] Charles Darwin, *The Life and Letters of Charles Darwin*, ed. Francis Darwin, vol. 1 (New York: Appleton, 1887), 278–279

Made in United States
North Haven, CT
04 November 2026

21733836R00055